Preschool

Counting & Coloring Workbook

Animals on the Farm

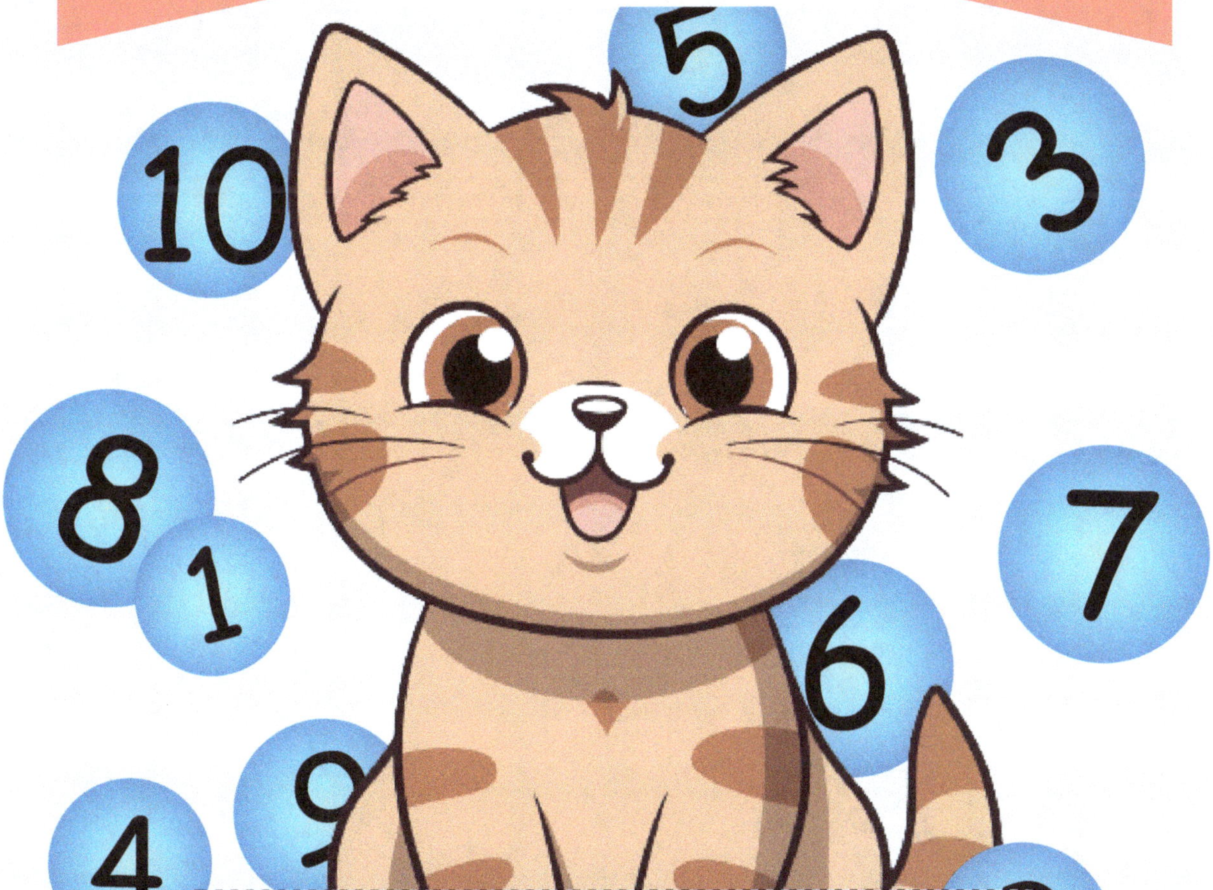

This belongs to:

cat

dog

chicken

llama

swan

goat

duck

donkey

turtle

COW

pig

chick

lamb

rooster

horse

mouse

rabbit

Let's Count!

Write the missing number on the empty eggs.

1   3

7   10

13

16   18

19

Numbers Tracing Worksheet

Trace the numbers below.

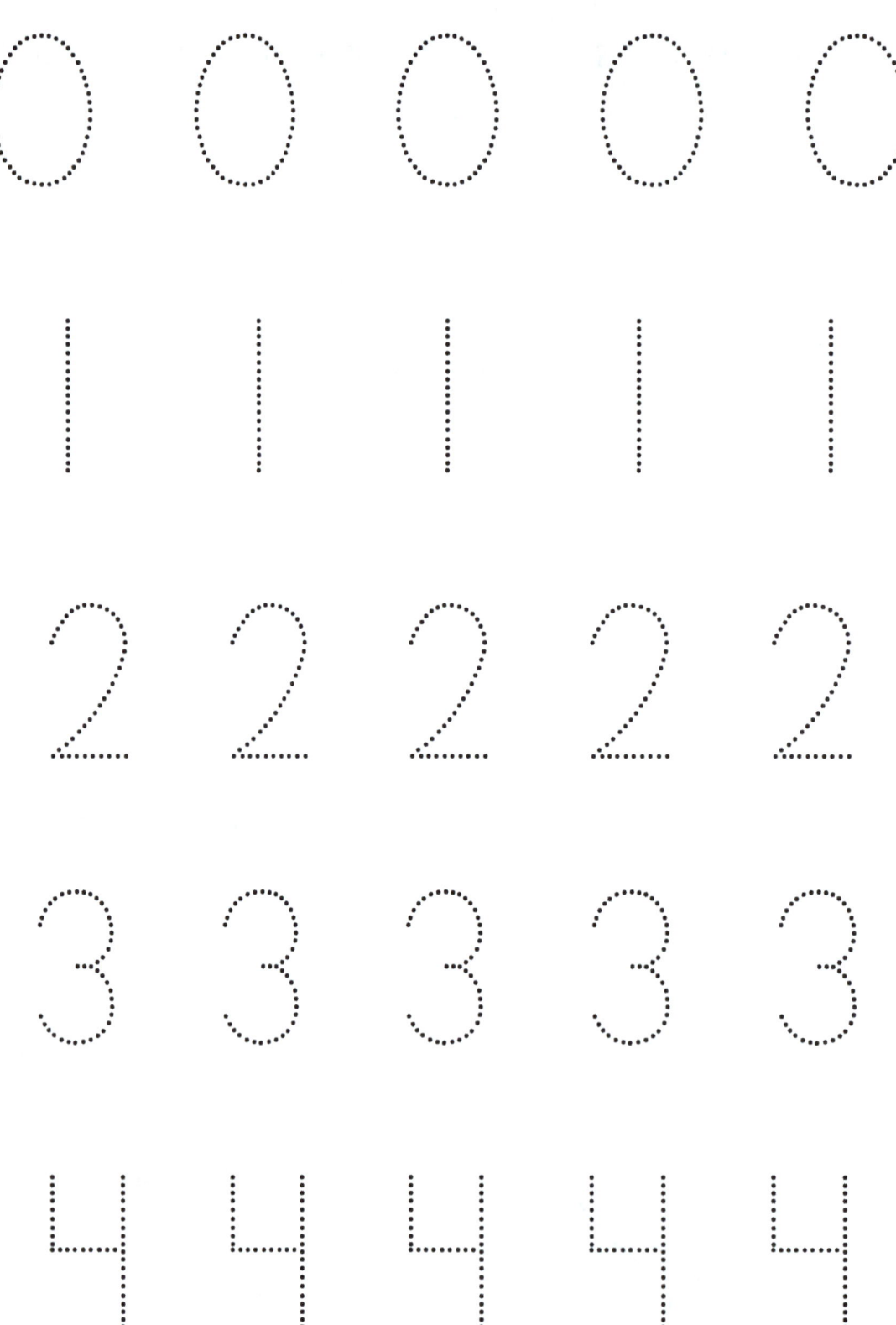

Numbers Tracing Worksheet

Trace the numbers below.

5 5 5 5 5

6 6 6 6 6

7 7 7 7 7

8 8 8 8 8

9 9 9 9 9

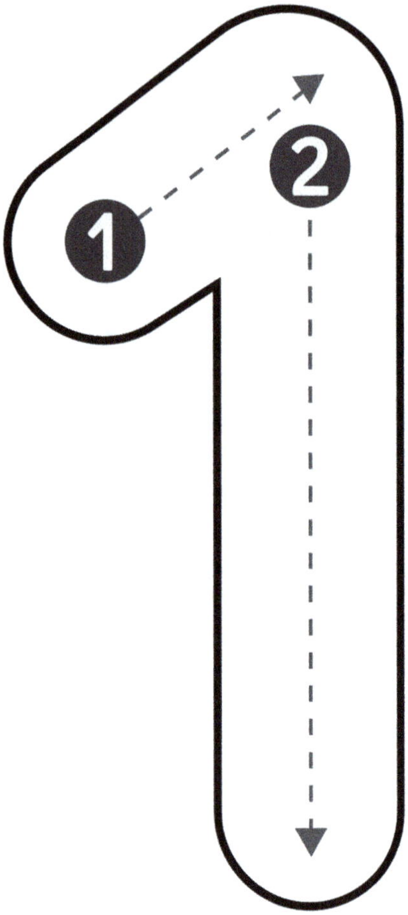

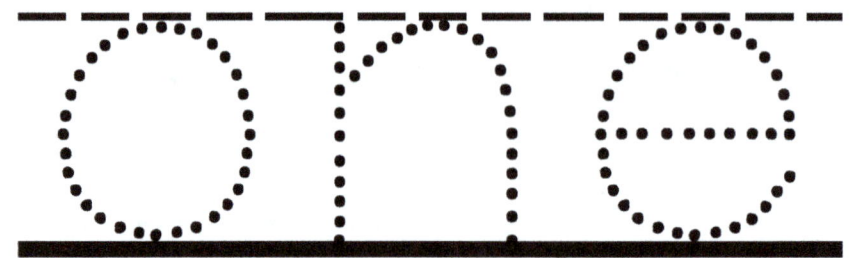

Time to count!

2

two

Time to count!

Time to count!

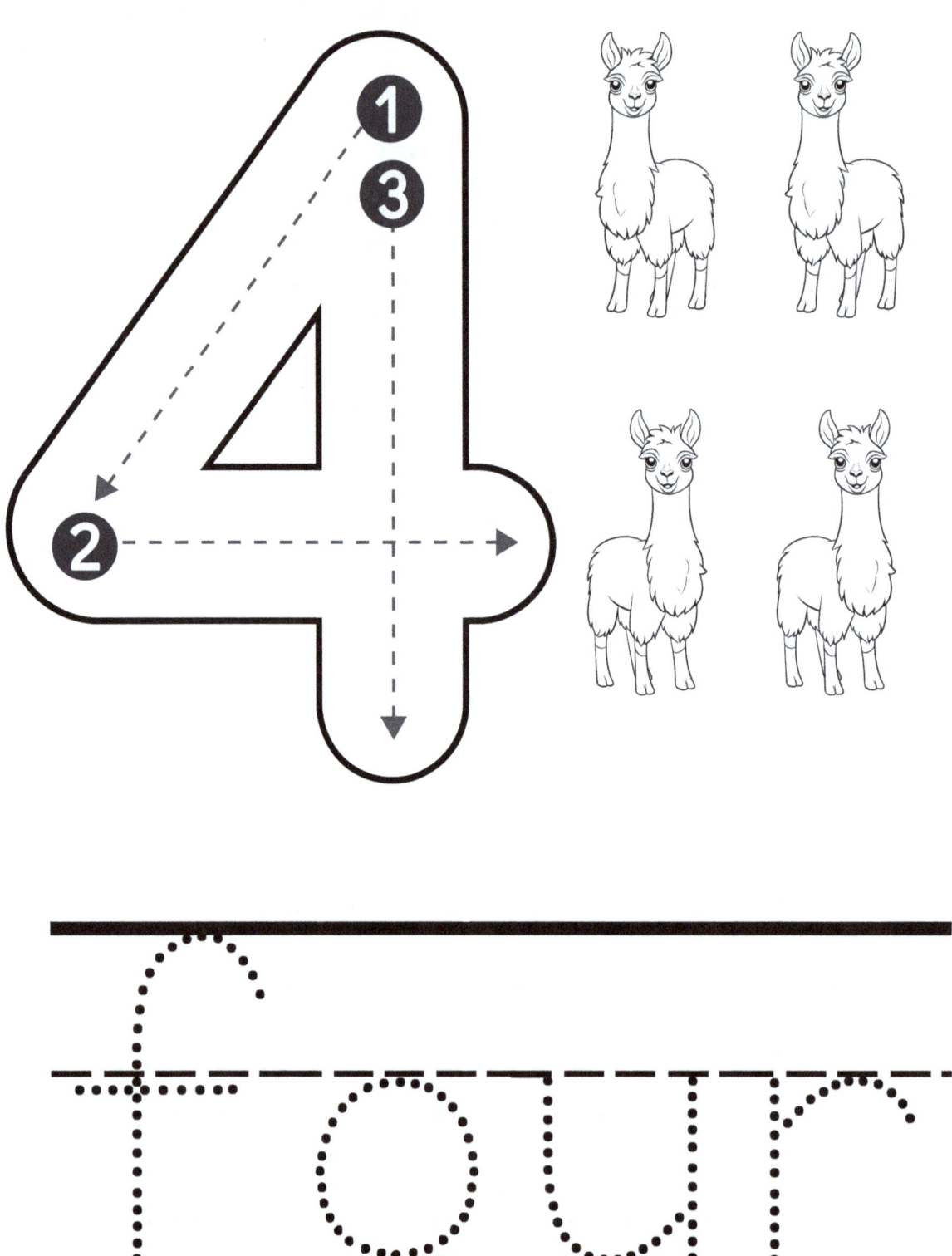

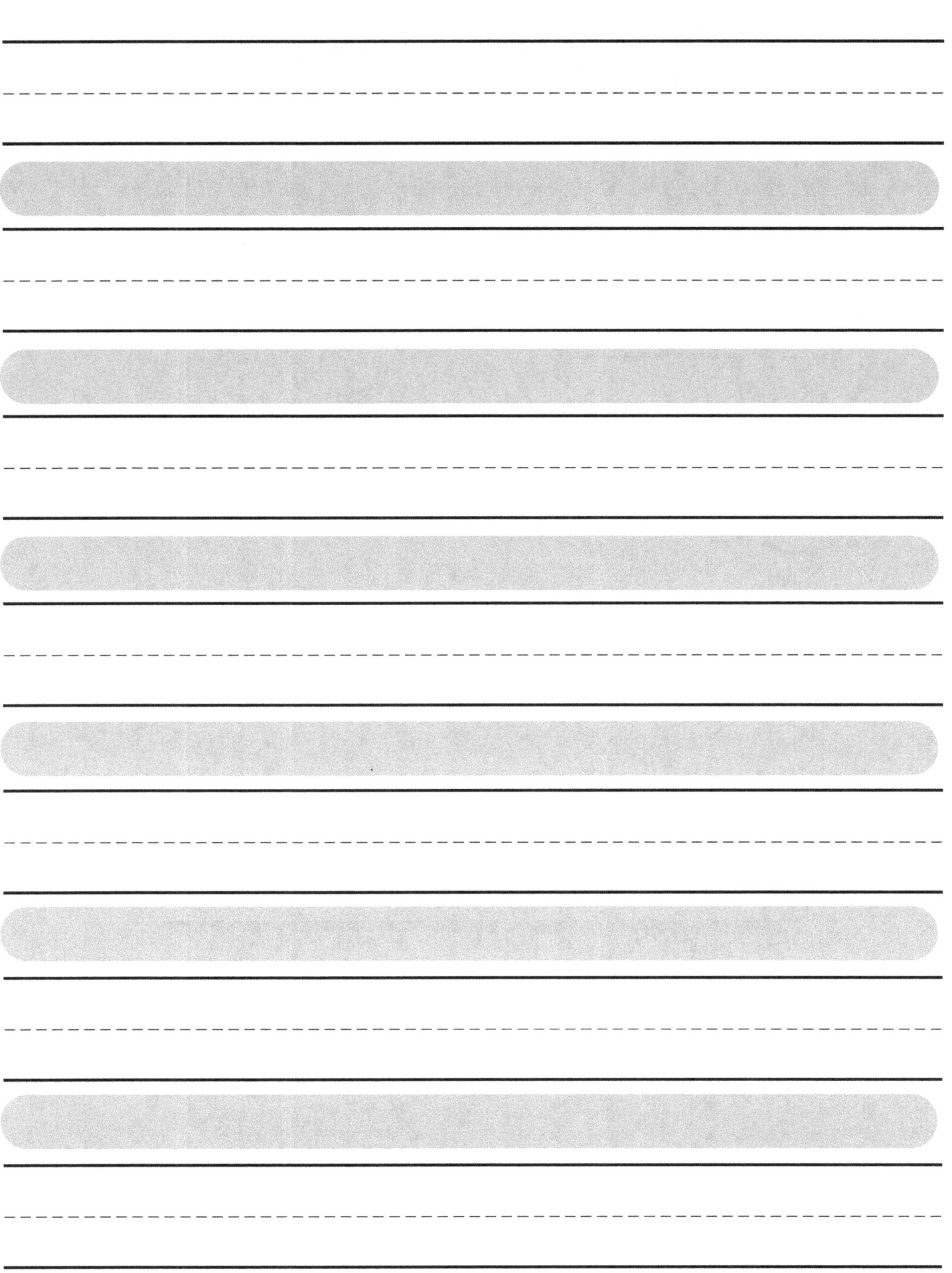

Time to count!

Time to count!

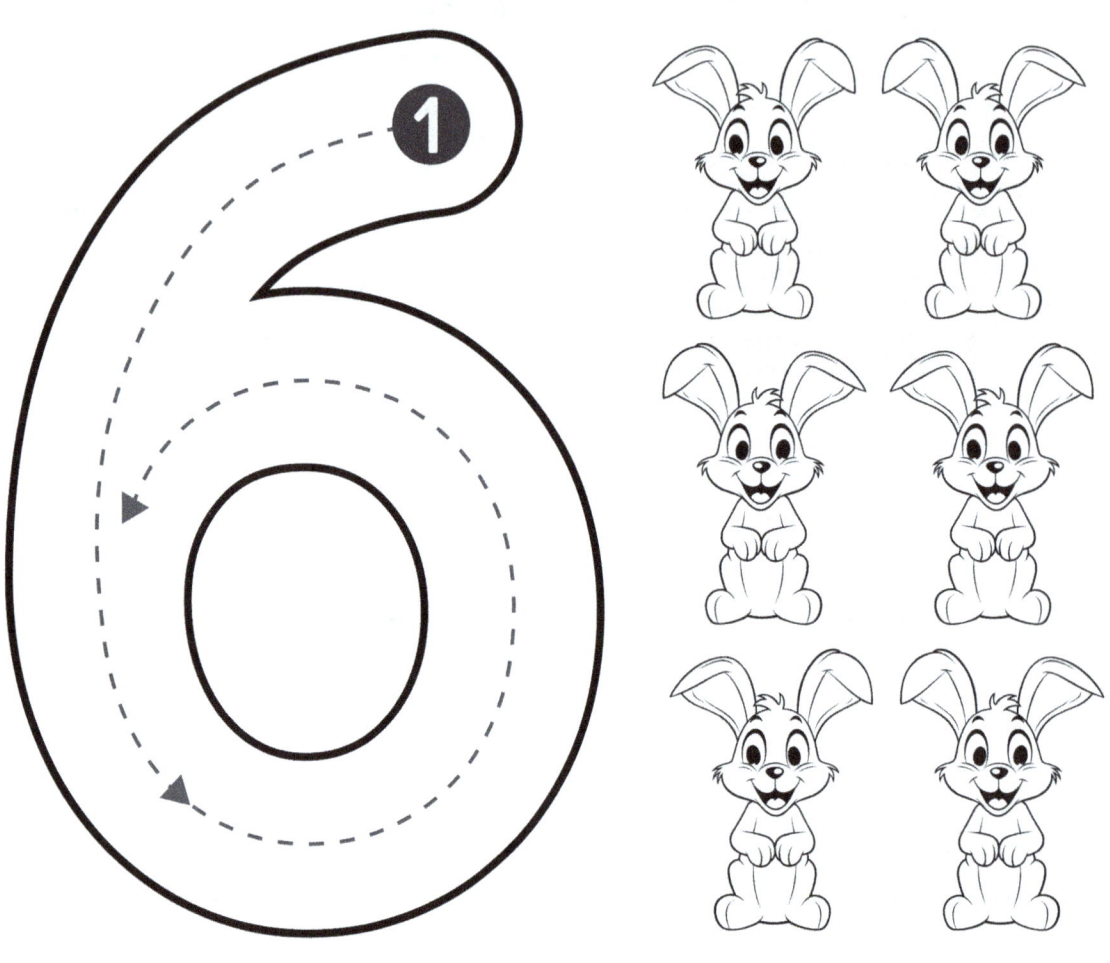

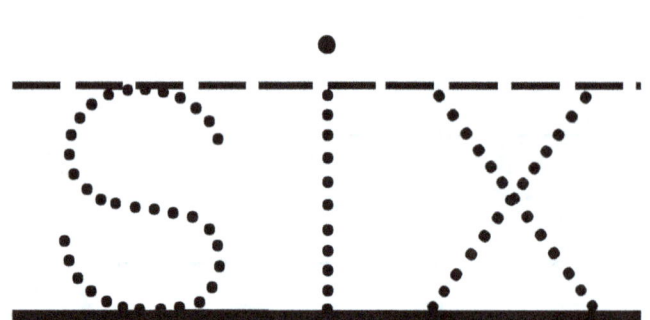

Time to count!

7

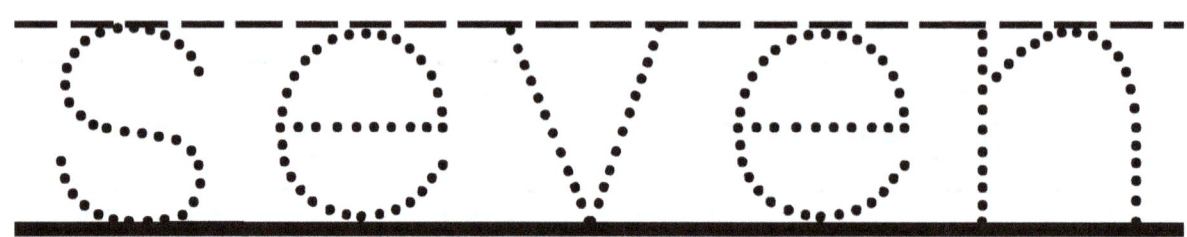

Time to count!

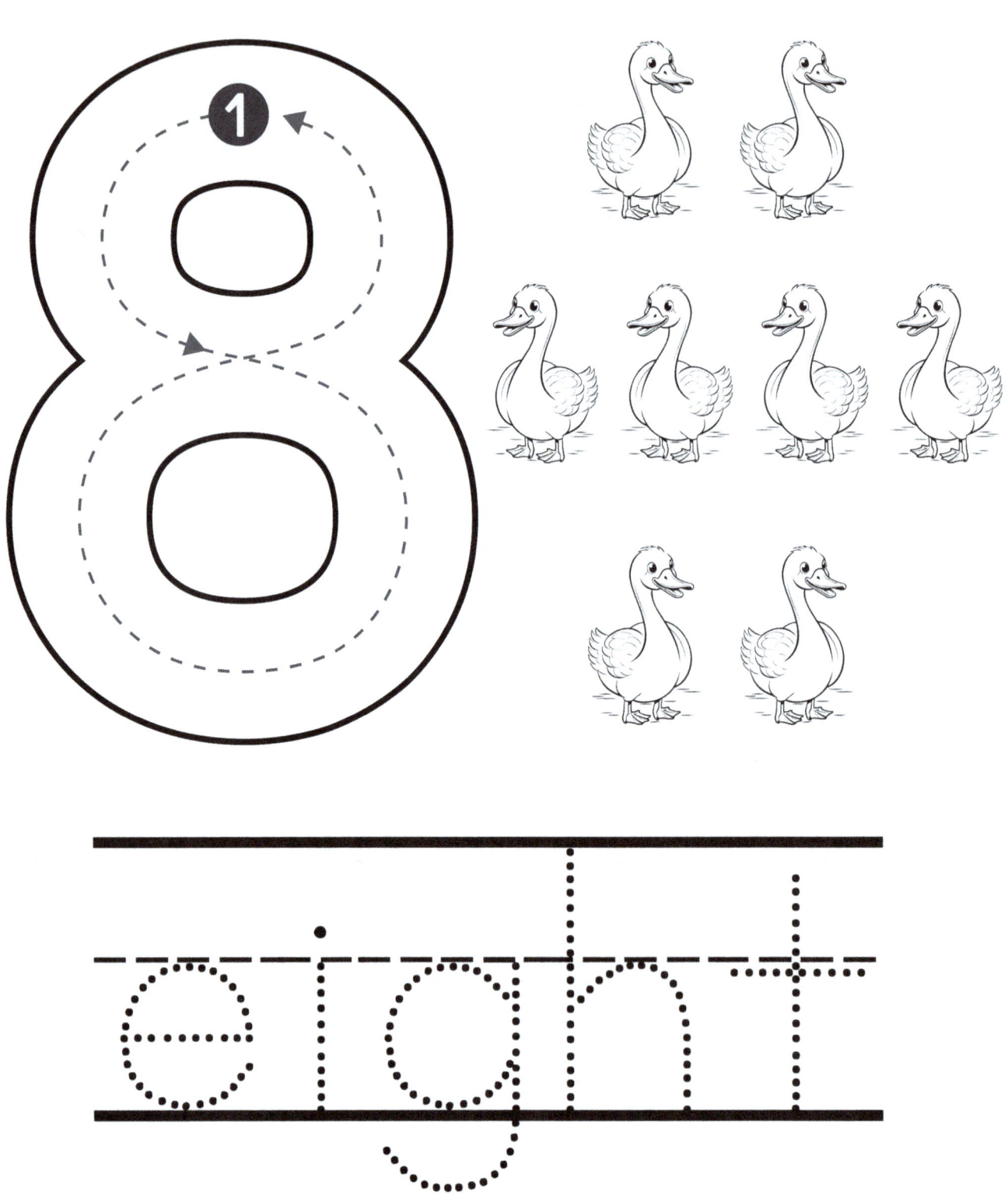

Time to count!

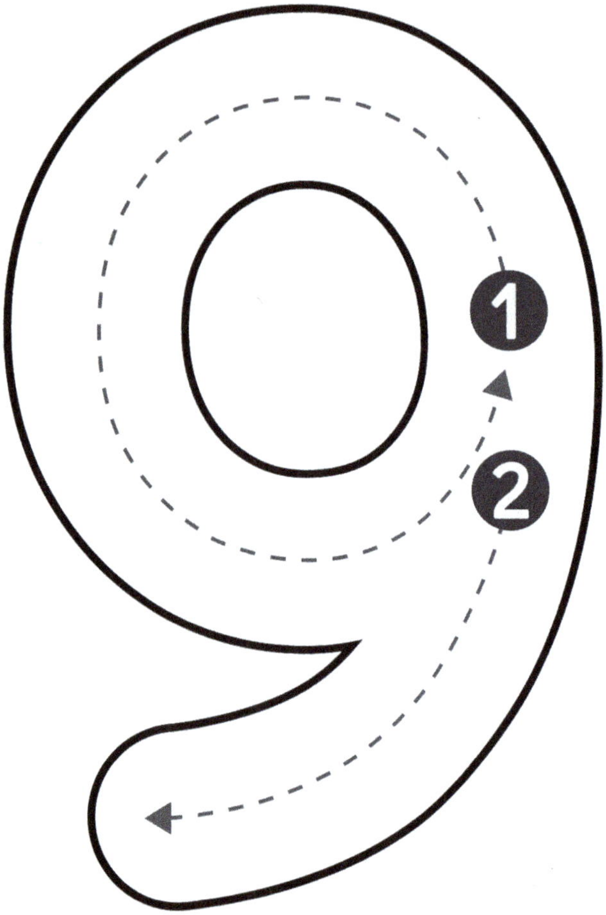

9

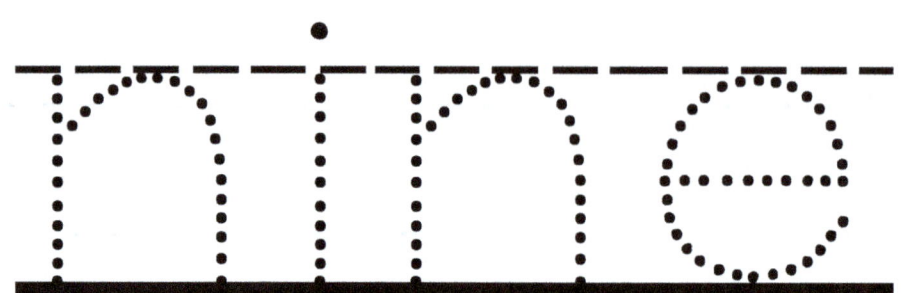

Time to count!

Counting animals

Count and circle the number of animals in each box.

6 5 7 8	3 4 5 6
1 2 3 4	9 8 10 2
9 6 10 2	5 2 9 8
10 3 8 2	1 10 5 9

How many?

Count the animals and write the numbers (1-10) in the boxes.

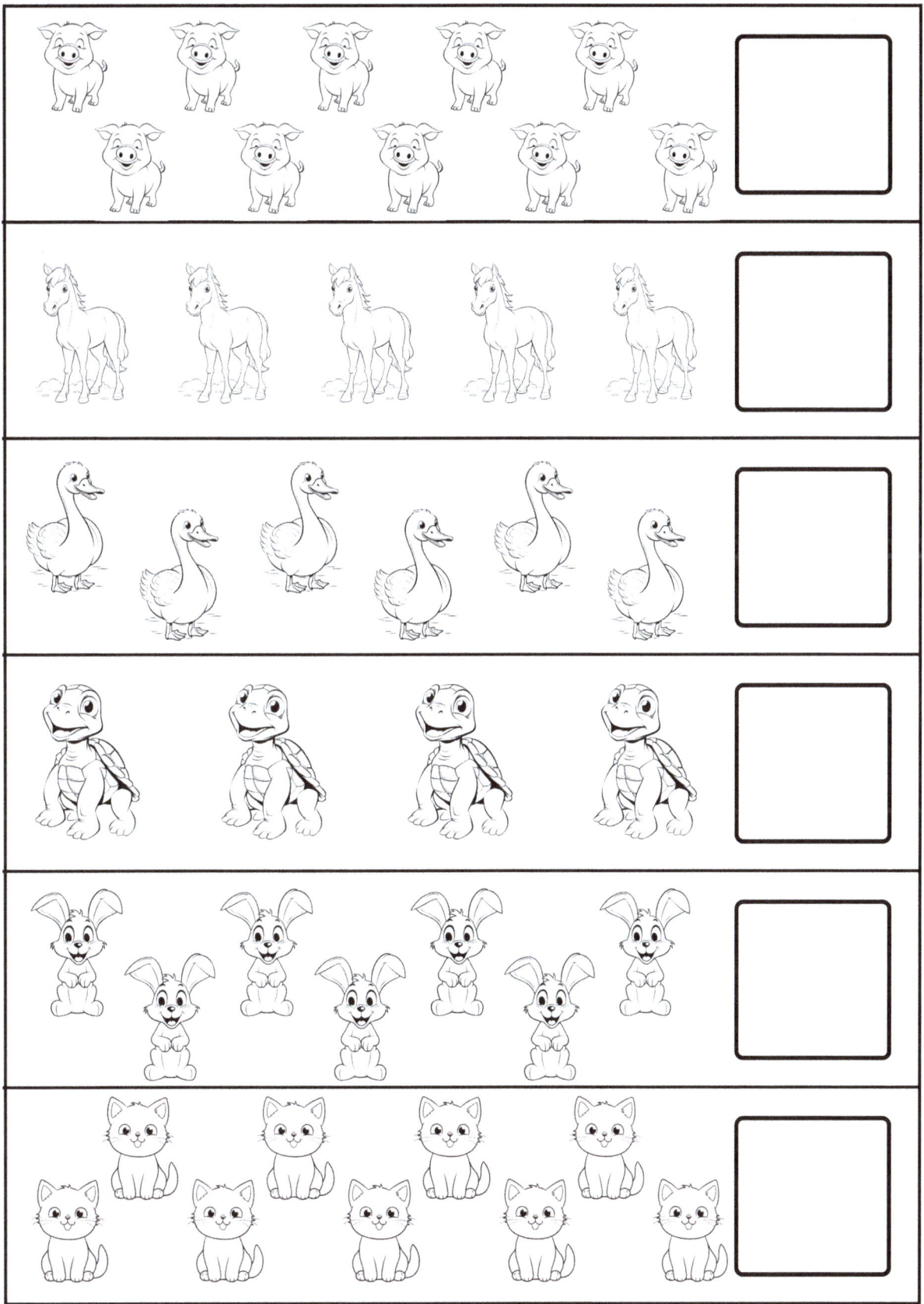

 # Let's count animals!

Count and write your answers in the chart below

Counting Birds

Count the number of objects, and circle the correct answer:

8   5   3   4

9   8   7   6

7   8   5   6

12   14   13   11

Join the numbers!

1   1

Trace

Color all the boxes with the number 1

1	1	4
1	3	4
1	7	1
9	5	4

one

Join the numbers!

Color all the boxes with the number 2

1	1	2
2	3	4
1	2	1
7	6	2

Trace

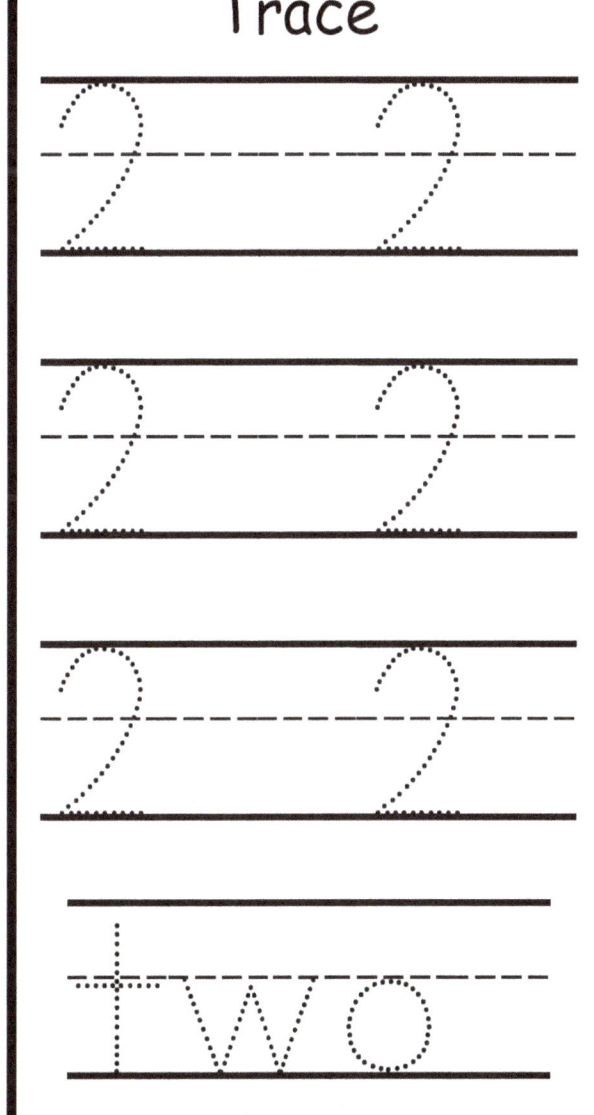

two

Join the numbers!

3

3

Color all the boxes with the number 3

2	4	**3**
3	3	3
1	7	1
9	3	3

Trace

3 — 3

3 — 3

3 — 3

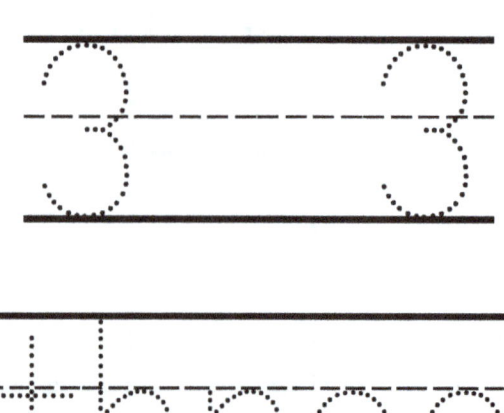

Join the numbers!

4   4

Color all the boxes with the number 4

5	7	4
4	3	8
1	2	1
4	5	4

Trace

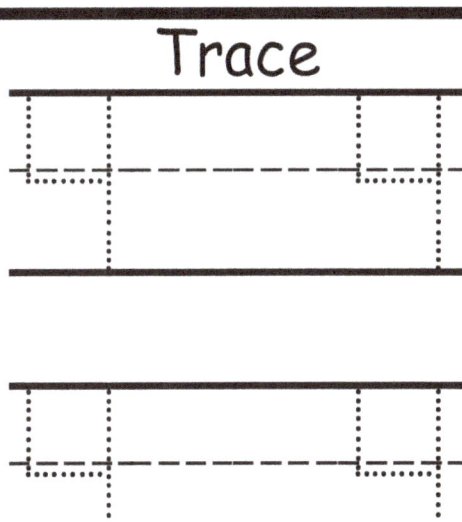

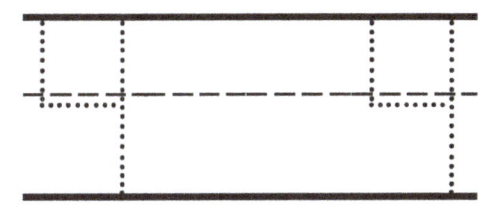

four

Join the numbers!

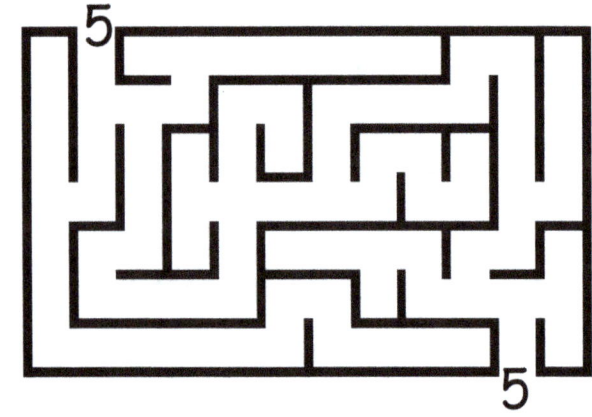

Color all the boxes with the number 5

5	1	5
3	5	5
7	7	1
9	5	6

Trace

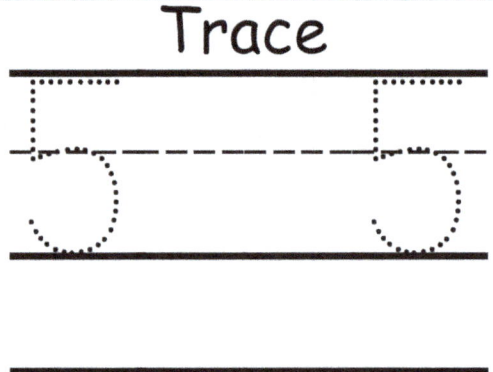

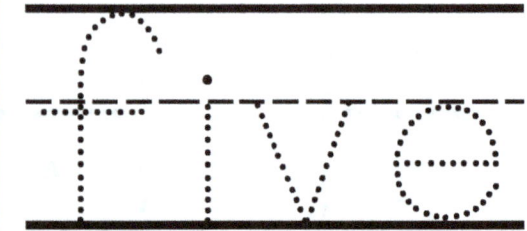

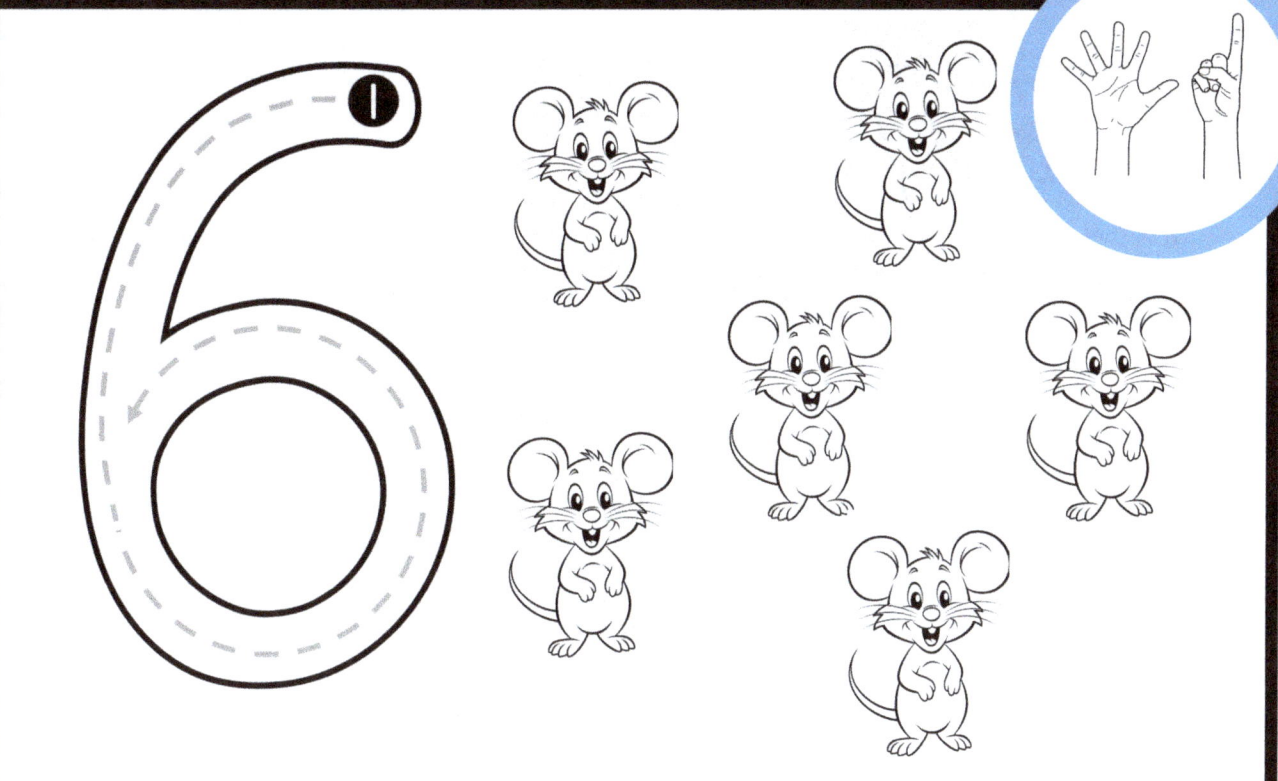

Join the numbers!

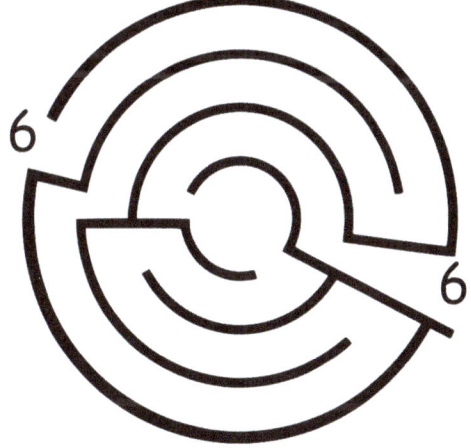

Color all the boxes with the number 6

6	6	6
9	2	4
2	6	1
6	5	6

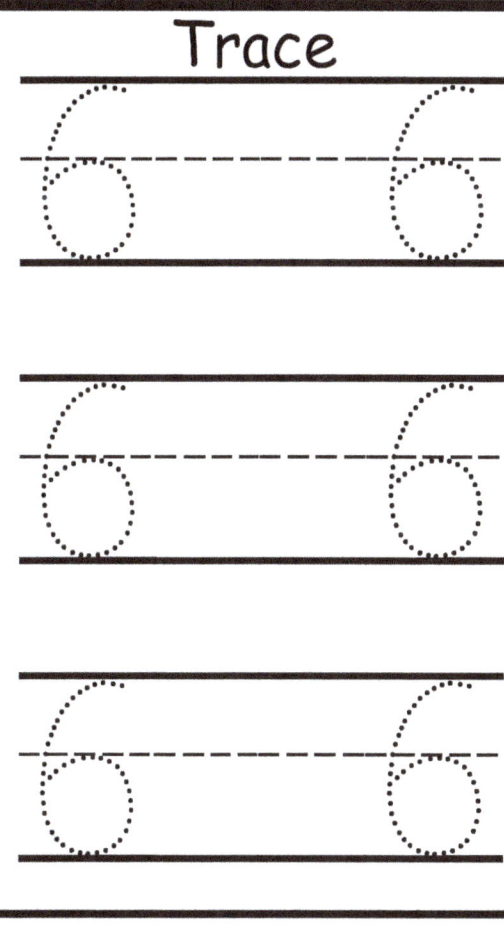

Trace

7

Join the numbers!

7

7

Color all the boxes with the number 7

2	1	7
7	7	5
1	7	1
7	5	2

Trace

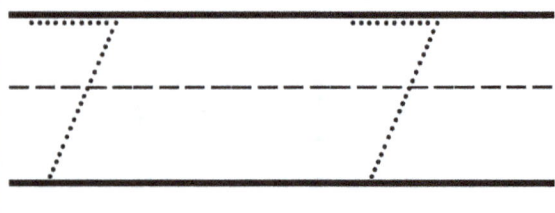

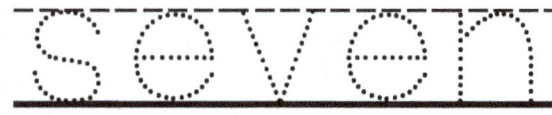
seven

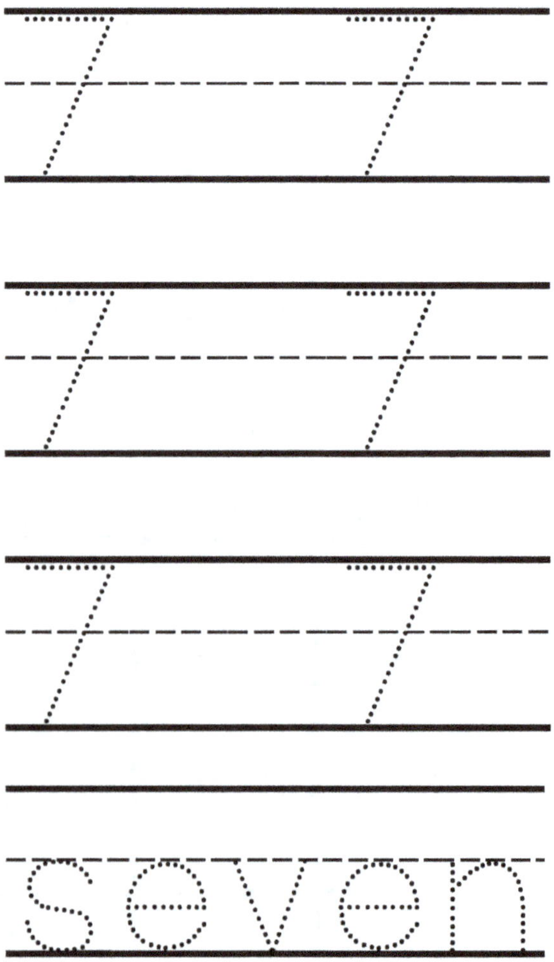

Join the numbers!

8

8

Color all the boxes with the number 8

8	8	8
1	3	4
8	7	7
3	8	4

Trace

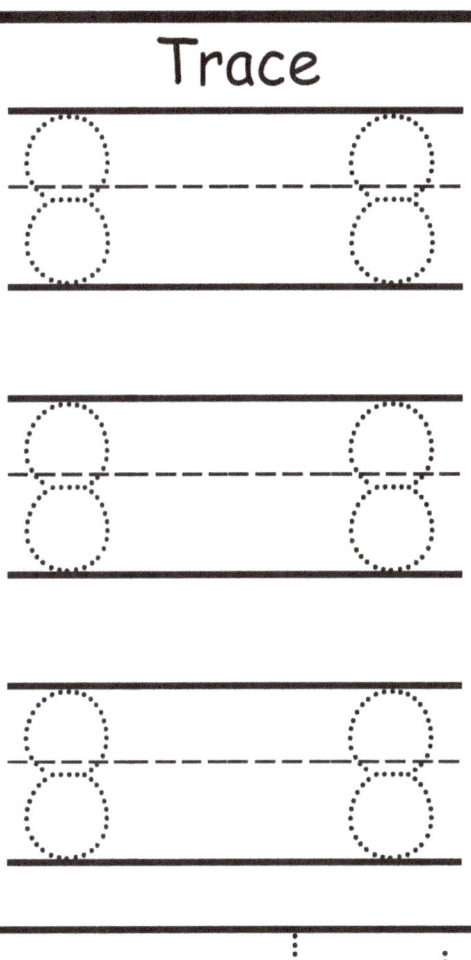

eight

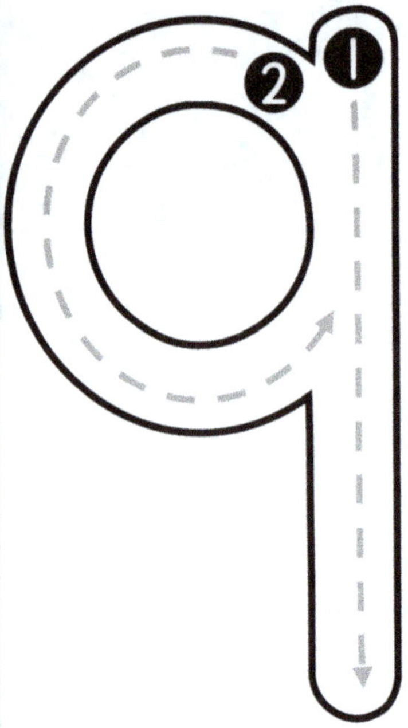

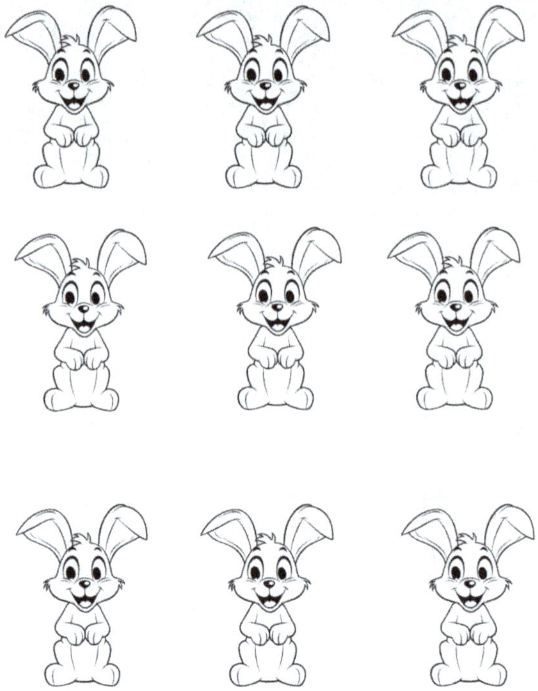

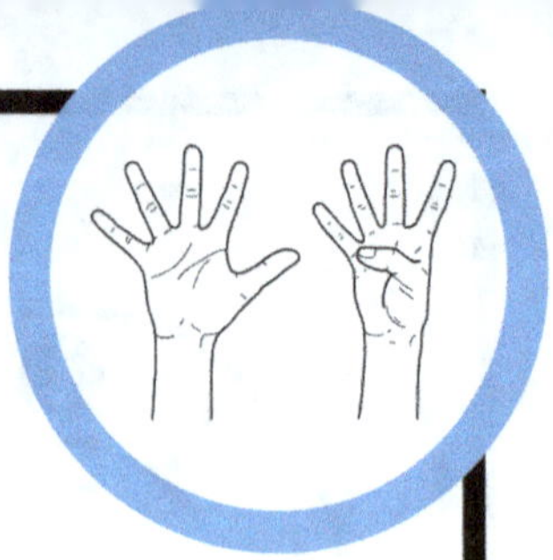

Join the numbers!

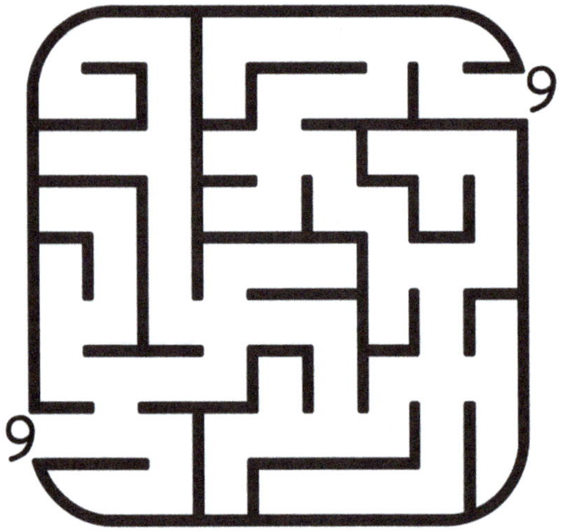

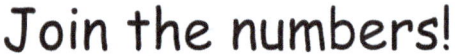

Color all the boxes with the number 9

9	1	9
1	9	4
2	7	9
9	5	4

Trace

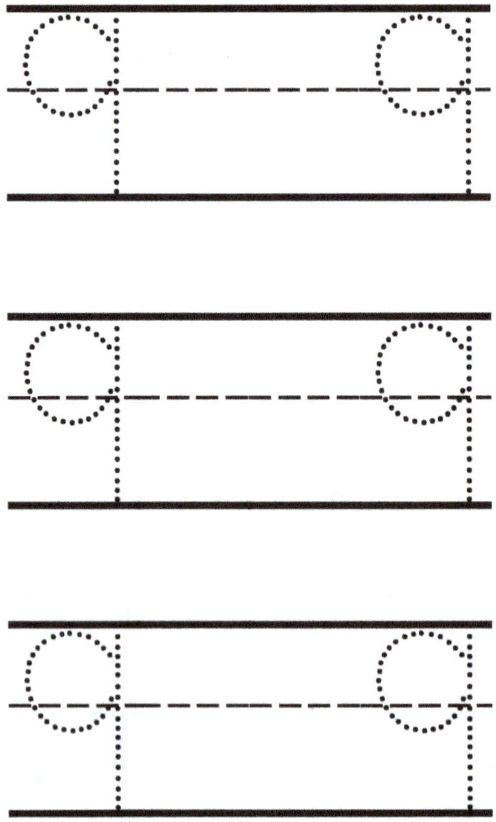

nine

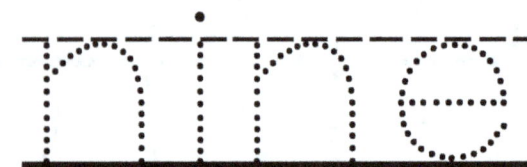

Join the numbers!

10

10

10

Color all the boxes with the number 10

10	9	8
5	2	9
9	10	1
10	5	10

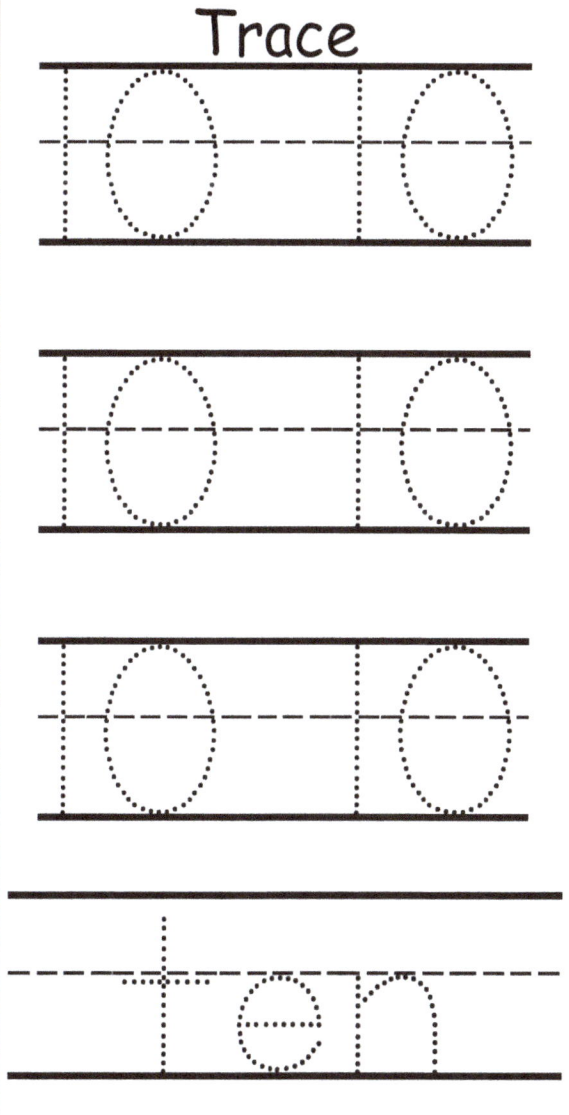

Trace

Number Order

Fill in the missing number in the sequence.

	2	3		

		8		10

11				15

16			19	

Number Order

Fill in the missing number in the sequence.

	2			5

6		8		

	12	13		

			19	20

Number Order

Fill in the missing number in the sequence.

| 1 | | 3 | | |

| | 7 | 8 | | |

| | 12 | | | 15 |

| 16 | | | | 20 |

Number Order

Fill in the missing number in the sequence.

1				5

6				10

11				15

16				20

Number Order

Fill in the missing number in the sequence.

| 1 | | | 4 | |

| 6 | | | 9 | |

| | 12 | | | 15 |

| | | 18 | | 20 |

Number Tracing

| 0 zero |
| 1 one |
| 2 two |
| 3 three |
| 4 four |
| 5 five |
| 6 six |
| 7 seven |
| 8 eight |
| 9 nine |
| 10 ten |

Numbers to 10

Color the amount of circles represented by each number:

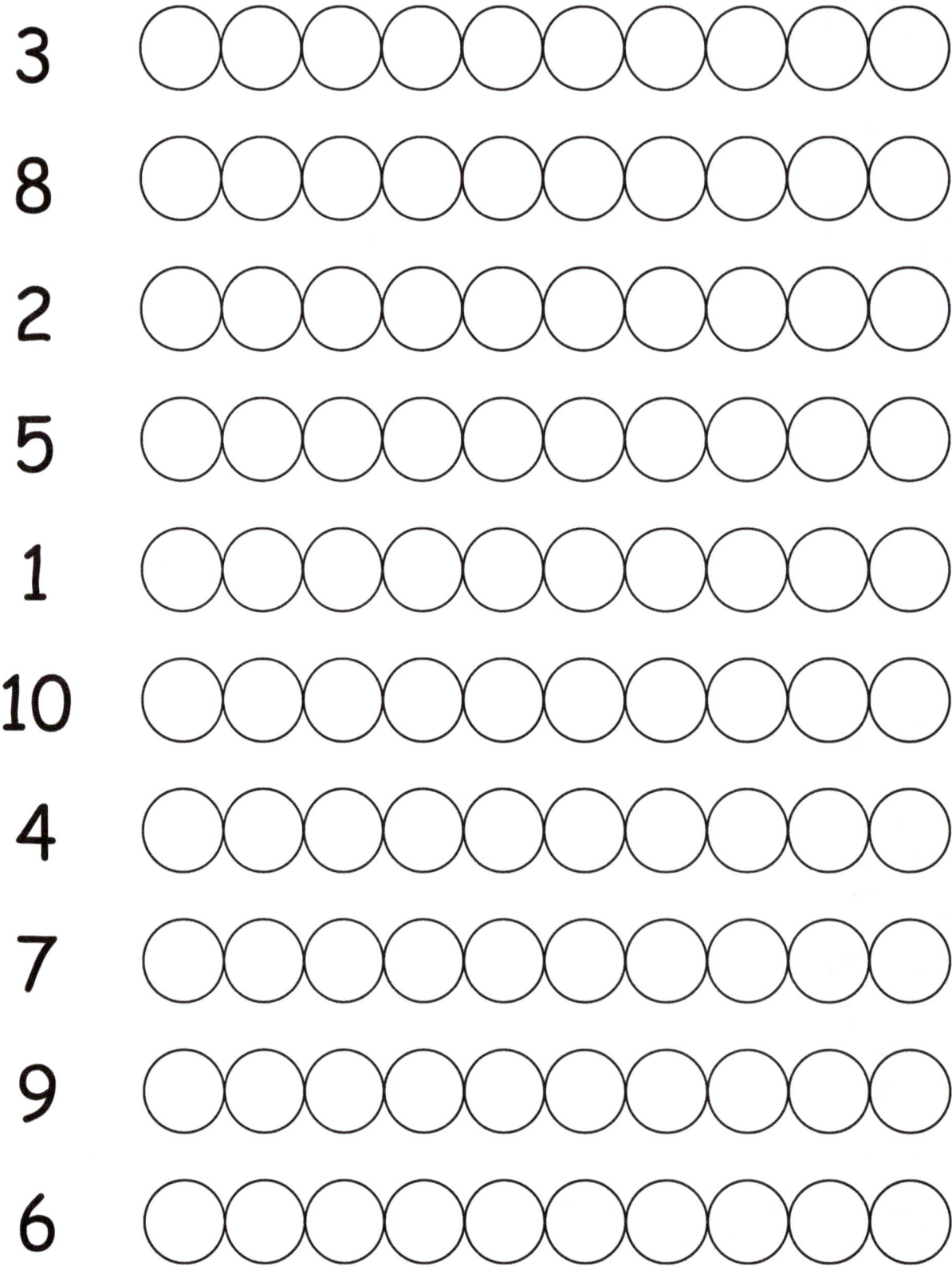

3

8

2

5

1

10

4

7

9

6